AP Biology Flashcard Quicklet

Flashcards in a Book for Biology Students

Paul Sanghera, Ph.D.

AP Biology Flashcard Quicklet: Flashcards in a Book for Biology Students

Published by
Infonential, Inc.
A California Corporation.
http://www.infonentialinc.com
email: info@infonentialinc.com

ISBN-10: 0-9791797-7-7
ISBN-13: 978-0-9791797-7-8

This publication is designed to provide accurate and authoritative information on the covered subject matter. It can be used by the students of biotechnology in conjunction with a textbook for a quick review of the subject matter. However, it is sold with the understanding that the publisher is not engaged in offering technical, legal, or other professional service. If such assistance is required, the service of a competent professional should be sought.

Warning and Disclaimer
The information published in this book has been obtained by Infonential, Inc. from sources believed to be reliable. We have made our best effort to make this material as comprehensive within the scope of this book and as accurate as possible. However, because of possible errors such as human, mechanical, or technical, the publisher and the author does not guarantee the accuracy, adequacy, or completeness of any information and are not responsible for any errors or omissions and the results obtained from such information.

To
Rosalind Franklin, James Watson, and Francis Crick
For discovering the structure of

DNA

Image: courtesy of www.ukcoinpics.co.uk

Table of Contents

The Quicklet Book Series

The purpose of a *quicklet* book is to present the required information on a subject within a well defined scope by focusing on the customer needs.

The lack of information is the problem of the past ages. The problem of the current age, the information age, is that we have too much information and too little time to absorb it. We are being bombarded with all kinds of needed and (mostly) un-needed information from all sides. This situation has created the need for information products that contain well defined, precise, to the point, quick, and "just what the customer needs" information. The *Quicklet Book Series* is a response to this market need. In other words, a *quicklet* is based on the philosophy that in this fast paced information age, nobody has the time to prowl through web pages or struggle with a 700 pages long book to find just that little piece of needed information.

Bottom line: A *quicklet* offers maximum learning in minimum time. No fluff, just the nuggets of the needed information presented in an easy to understand format.

About the Author

Dr. Paul Sanghera, an educator, scientist, technologist, and an entrepreneur, has a diverse background in multiple fields including physics, chemistry, biology, computer science, and math. He holds a Master degree in Computer Science from Cornell University, a Ph.D. in Physics from Carleton University, and a B.Sc. with triple major: physics, chemistry, and math. He has taught science and technology courses all across the world including San Jose State University and Brooks College. Dr. Sanghera has been involved in educational programs and research projects in biotechnology. He has authored and co-authored more than 100 research papers published in well reputed European and American research journals.

As a technology manager, Dr. Sanghera has been at the ground floor of several technology startups. He is the author of several best selling books in the fields of science, technology, and project management. He lives in Silicon Valley, California, where he currently serves as Assistant Professor at California Institute of Nanotechnology.

About This Book

Dr. Paul Sanghera offers a taste of the discipline of biology for beginners by presenting more than 325 flashcards in this easily portable book. Primarily designed to help you prepare for the advanced placement (AP) biology exam, this book is also a quick introduction to (or overview of) biology for beginners. Although the book is self contained within its scope, it assumes that you have already studied the material from a reference such as an AP Biology Exam Study Guide or an introductory biology book.

These flashcards offer a quick and logical review of your preparation for the AP Biology Exam, and are designed to improve your information recall capability and response speed during the exam.

Special features:

♦ All essential concepts, terms, and processes are covered.

♦ The depth and style of the coverage makes these flashcards indexes into your memory so that if you go through these flash cards after reading a study guide, it's almost equivalent to going through the study guide once again, only in much less time.

♦ The flashcards are largely self-contained and no reference to any other book is made.

How to use this book: Read the front of a page and answer it. Then look at the back of the page to check if your answer was correct.

Welcome to the discipline of biology.

Enjoy.

Q1

What is science?

Q2

In which way science is a process?

A1

Science is the body of knowledge about the natural world organized in a rational and verifiable way. The word *science* has its origin in a Latin word that means "to know". Science is organized into different fields or branches such as physics and chemistry. Mathematics (math in short) is considered to be the language of science, as it is used to express science, for example, in formulas.

A2

Science is a way of knowing. It involves discovery, which can use a process of inductive reasoning. It can also use hypothesis testing, which is also a process.

Q3

What is biology?

Q4

What are organisms and microorganisms?

A3

Biology is the branch of science that studies life. The word biology has a Greek origin: *bio* means life and *logos* means knowledge. Biology is also referred to as biological sciences because it has a broad spectrum of fields including botany (study of plants), zoology (study of animas), cellular biology (study of basic building blocks of life), microbiology (study of microorganisms), and ecology (study of interrelations among organisms).

A4

microorganism An organism that is too small to be seen by human eye

organism A living adaptive system of organs with organs interacting in such a way that they work as a single whole.

organ A collection of tissues that perform a set of functions.

tissue A collection of interconnected cells within an organism with each cell performing a similar function.

cell The smallest metabolically functional unit of life largely made of protein and DNA.

adaptive system A system capable of learning from experience, and changing accordingly.

Q5

Biology is based on which two branches of science?

Q6

What is chemistry?

A5

Physics and chemistry

The foundations of biology are based on physics and chemistry. In other words, biology applies the laws of physics and chemistry to living things.

A6

Chemistry is the branch of science that deals with the composition, properties, and interactions (reactions) of matter especially at the atomic and molecular levels.

Q7

What is biochemistry?

Q8

What is biotechnology?

A7

The study of the fundamental chemistry of living things, that is, organisms

A8

Any technology that uses biological systems, living organisms, or parts of living organisms to make or modify products, processes, or applications for specific uses.

Q9

What are the characteristics of living things?

Q10

What is the complexity of organization in organisms?

A9

- **Complexity of organization**
- **Ecology**
- **Evolution**
- **Growth**
- **Metabolism**
- **Reprodction**
- **Responsiveness**

A10

Complexity of organization refers to the complex and sophisticated structure of organisms. For example, an organism is composed of organs, which are composed of tissues. A tissue in turn is composed of cells with each cell performing a similar function. Each cell is composed of molecules.

Q11

What is ecology?

Q12

What is evolution?

A11

Ecology is the study of releationships among different organisms and the realtionships of organisms with their environment.

A12

Evolution may be defined as a change in the genetic composition of a genetic population from generation to generation as a result of processes such as natural selection and mutation. This continuous change ultimately results in the development of new species.

Q13

What is the growth chaarctersitic of a living organism?

Q14

What is metabolism?

A13

The ability to grow; this is the ability of an organism to take in the appropriate material from its environment and convert it into a part of its own structure. Eating and digesting food is an example.

A14

The set of all biochemical processes (chemical reactions) occurring in a living cell of an orgnaism. These processes enable cells to maintain their structure, grow, reproduce, and respond to their environemnt. In this way, metabolism provides the basis of life. It involves two steps: converting the consumed material into energy, called catabolism, and using that energy to build cell commpnents, called anabolism.

Q15

What is reproduction?

Q16

What are the two types of reproduction?

A15

A fundamental feature of life that enables organisms to produce new organisms. All organisms exist as a result of reproduction.

A16

Sexual Two individual organisms (parents) participate in producing the new individual. Humans reproduce through sexual reproduction.

Asexual An individual (one parent) can reproduce without the participation of another individual. Most plants have the ability for asexual reproduction.

Q17

What is responsiveness?

Q18

What is a scientific method?

A17

The ability of living things to respond to the external stimuli (the stimuli in the external environment). Examples of stimuli are light, sound, heat, and contact. Humans use five senses (seeing, hearing, smell, taste, and touch) to detect the stimuli.

A18

A process to obtain scientific knowledge. It has the following three general steps:

1. Observe a phenomenon and state the problem

2. Formulate hypothesis that explains observation. Make predictions based on the hypothesis

3. Test the hypothesis and predictions through experiments

Q19

What is the difference between a theory and a law?

Q20

What is physics?

A19

A hypothesis is a logical and verifiable explanation of a phenonmenon or observations that makes some predictions. A hypothesis that passes the experimental tests is usually called a theory. A well tested and widely accepted theory is usually accepted as a law among the scientific community.

A20

Physics is that branch (or discipline) of science that deals with understanding the universe and the systems in the universe in terms of fundamental constituents of matter (such as atoms, electrons, and quarks) and the interactions between those constituents.

Q1

What is chemistry?

Q2

What is an element?

A1

Chemistry is the branch of science that deals with the composition, properties, and interactions (reactions) of matter especially at the atomic and molecular levels.

A2

An element, also called chemical element, is a type of atom defined by the number of protons in its nucleus, called atomic number. It also refers to a pure chemical substance composed of atoms with the same atomic number. This brings us to another definition of element:

An element is a substance that cannot be chemically decomposed.

All living (and non living) things on earth are composed of elements. As of 2007, 117 elements are known out of which 94 exist naturally. Some examples are carbon, chlorine, hydrogen, nitrogen, and oxygen.

Q3

What is an atom?

Q4

What are atomic orbitals?

A3

An atom is the smallest particle that characterizes an element. The word atom has a Greek origin in the word *atomos* means indivisible. An atom is composed of a nucleus and electrons, negatively charged particles, which revolve around the nucleus. The nucleus is composed of two kinds of particles called protons and neutrons. Protons are positively charged, whereas neutrons are electrically neutral. None of electron, proton, and neutron individually represents an element, whereas an atom does.

A4

These are the states in which electrons in an atom can be found. From a classical perspective, electrons in an atom are considered to be orbiting the nucleus of the atom. Orbitals can be expressed in shells and sub-shells, also called energy levels and energy sub-levels correspondingly. A shell is denoted by n, an integer. A shell contains sub-shells, which are written in the following form:

$$nx^y$$

```
where n is the shell number (an integer), x is the
label for the subshell (s, p, d, f, g...), and y is
the number of electrons in the subshell.
```

Q5

How are subshells distributed in shells?

Q6

How are electrons in an atom distributed in shells and subshells?

A5

First shell ($n=1$) can contain only subshell s.

Second shell ($n=2$), can contain subshells s and p.

Third subshell ($n=3$) can contain subshells s, p, and d.

And so on...

The shell number n is also called the principal quantum number.

A6

A shell can contain up to $2n^2$ electrons. For example shell number 1 can contain up to 2 electrons and shell number 2 can contain up to 8 electrons.

A subshell can contain a maximum number of electrons given by

$4l+2$

where l is a whole number: 0, 1, 2, 3, 4...

These values of l are symbolized as s, p, d, f, g...

Therefore, the subshell s can contain up to 2 electrons, subshell p up to 6 electrons, subshell d upto 10 electrons, and so on.

Q7

What is a molecule?

Q8

What is a chemical compound?

A7

A molecule is a group of two or more atoms organized in a definite arrangement held together by chemical bonds. The atoms in a molecule may be of the same type as in O_2 or of different types as in H_2O.

A8

A chemical compound is a substance composed of two or more different chemical elements bonded together. For example, water (H_2O) is a compound with each molecule consisting of two atoms of hydrogen bonded to one atom of oxygen.

Q9

What is an organic compound?

Q10

What are chemical bonds?

A9

Chemical compounds of living things are called organic compounds. The molecules of organic compounds contain carbon. Some examples are benzene (C_6H_6), ethane (C_2H_6), and Methane (CH_4).

A10

The bonds based on attractive force that binds together the atoms in a molecule and molecules in a substance. These bonds are based on the attractive electric force between the negatively charged electrons and positively charged protons in atoms.

Q11

What are some types of chemical bonds?

Q12

What are acids and bases?

A11

Covalent bond. A bond between two atoms formed by sharing one or more pairs of electrons. For example, in a methane molecule (CH_4), the carbon atom shares a pair of electrons with each of the four hydrogen atoms.

Ionic bond. A bond formed between two atoms when one or more electrons transfer from one atom to the other atom making one atom positively charged and the other atom negatively charged, and thereby causing the attractive electric force between them. For example, in ordinary table salt, that is, sodium chloride ($NaCl$), the bond between a sodium atom and a chlorine atom is an ionic bond because sodium loses one electron to chlorine.

Hydrogen bond. This bond id formed due to the attraction of a hydrogen nucleus (proton) to a negatively charged atom such as oxygen or nitrogen in the same or neighboring molecules. Example: water molecules.

A12

An acid is a chemical compound that releases a hydrogen ion (H^+) when mixed with another compound. An example is acetic acid (CH_3COOH) that gives vinegar the sour taste.
A basis (or base) is a chemical compound that can accept H^+. Note that this is opposite to an acid. An example is sodium hydroxide ($NaOH$).

Q13

What are carbohydrates?

Q14

What are different types of carbohydrates?

A13

A carbohydrate is an organic compound that consists of carbon, hydrogen, and oxygen. Carbohydrates serve as sources of energy for organisms and some carbohydrates also serve as structural materials. Some examples are simple sugar, starch, and cellulose.

Carbohydrates are also called *saccharides*.

A14

Monosaccharide. A carbohydrate composed of single molecules, such as glucose ($C_6H_{12}O_6$).

Disaccharide. A carbohydrate composed of two molecules (or two monosaccharides) such as sucrose. Disaccharides are the simplest polysaccharides.

Polysaccharides. Carbohydrates composed of multiple monosaccharides. An example is starch.

Chemistry of Life

Q15
What is a chemical substance?

Q16
What is a chemical reaction?

A15

A material with a well defined definite chemical composition. For example, pure water is a chemical substance because it is composed of hydrogen and oxygen with a definite ratio of 2 to 1 regardless where the pure water comes from. A chemical substance can only be converted to another chemical substance through a process called chemical reaction.

A16

A chemical reaction is a process in which one or more chemical substances are converted into other chemical substances. The initial substances involved in a chemical reaction are called reactants, and the substances generated from the reaction are called products. During the reaction, the atoms of the substances rearrange themselves resulting in making or breaking chemical bonds due to the motion of electrons. Energy may be absorbed or released during a chemical reaction.

Q17

What is a chemical formula?

Q18

What is a chemical equation?

A17

A symbolic representation of a chemical substance that shows the number of atoms of elements composing a molecule of the substance. For example, consider the following chemical formula for methane:

CH_4

It shows that a molecule of methane contains one carbon atom and four hydrogen atoms bonded together.

A18

A chemical equation is a symbolic representation of a chemical reaction with the reactants on the left and the products on the right of an arrow or an equal sign. For example, consider the combustion of methane in oxygen, which generates carbon dioxide and water. This chemical reaction can be represented by the following equation:

$CH_4 + 2O_2 = CO_2 + 2H_2O$

The coefficient preceding a symbol (or formula) represents the quantity of the entity represented by the symbol. For example, $2O_2$ means 2 molecules of oxygen, that is, four atoms of oxygen. A chemical reaction conserves charge and the quantity of each element. That means the electric charge and the quantity of each element on both sides of the equation should be identical.

Q19

List some examples of carbohydrates.

Q20

What are lipids?

A19

Glucose. A monosaccharide with the formula $C_6H_{12}O_6$; Basic form of fuel (energy) in living things.

Sucrose. Table sugar, a disaccharide formed by linking two monosaccharides: glucose and fructose.

Starch. A type of polysaccharides formed by linking hundreds or thousands of glucose units. We satisfy our energy needs by consuming starches from corn, potatoes, rice, and wheat.

Glycogen. A polysaccharide composed of thousands of glucose units and is mostly used for storing glucose in animals and humans mainly in the liver.

Cellulose. A polysaccharide composed of glucose units and is primarily used as a structural carbohydrate. For example, it's a chief constituent of the cell walls of plants.

A20

A group of organic compounds that are fat-soluble but water insoluble (hydrophobic) and include fatty acids and their derivatives. There are several types of lipids supporting different functions in living organisms such as storing energy, serving as structural components of cell membranes, and constituting signaling molecules.

Q21

What are proteins?

Q22

What are nucleic acids?

A21

A group of organic compounds composed of chains of amino acids, which contain carbon, hydrogen, nitrogen, and oxygen. A protein is a relatively large organic compound that consists of amino acids arranged in a linear chain and joined together by linking the amine nitorogen of one amino acid and a carboxyl atom of another, a bond called a peptide bond. Proteins are fundamental components of all cells, the basic functional units of life.

A22

A nucleic acid, found in all living cells and viruses, is a complex biochemical molecule with a high molecular weight, and is composed of nucleotide chains that convey genetic information. The most common nucleic acids are deoxyribonucleic acid (DNA) and ribonucleic acid (RNA).

A nucleotide is a chemical compound that consists of a nitrogenous base, one or more phosphate groups, and a sugar (a kind of carbohydrate) molecule.

Q23
What is a peptide?

Q24
The link between two amino acids is called a:

A23

A molecule composed by linking various amino acids called alpha amino acids. A peptide is usually smaller than a protein. Many hormones and antibiotics are examples of peptides. Proteins are polypeptides, that is, long peptides.

A24

peptide bond

Q1
What is a cell?

Q2
Describe a brief history of the premises of cell theory.

A1

A cell is the smallest metabolically functional unit of life largely made of proteins and DNA. All living beings, animals (including human) and plants, are made of cells.

A2

♦ **Mid 1600s.** An Englishman with the name of Robert Hooke observed tiny compartments in the cork that he examined through the newly invented microscope. He referred to these units as cells.

♦ **Late 1600s.** Anton Van Leeuwenhock, a Dutch merchant, studied cells in plants and animals.

♦ **1838.** Mathias Schleiden, a German botanist, proposed that all plants are composed of cells.

♦ **1839.** Theodore Schwann, a colleague of Mathias Schleiden, proposed that all animals like all plants, are also composed of cells.

♦ **1858.** Rudolf Virchow, a biologist, proposed that all living things are made of cells and that all cells form from already existing cells.

Q3

What are the two types of cells?

Q4

What is a cell composed of?

A3

Prokaryotic. These cells are usually singletons, that is, found in organisms composed of a single cell. An example is bacteria. These cells have no cell nucleus or membrane-bound compartments.

Eukaryotic. These cells are mostly found in multicellular organisms. A eukaryotic cell could be 1000 times larger in volume than a prokaryotic cell. These cells have cell nucleus and membrane-bound compartments.

A4

♦ **Nucleus.** Primarily composed of protein and DNA. DNA in a cell is organized into units called chromosomes. Functional segements of chromosomes are called genes. Only Eukaryotic cells have nucleus.

♦ **Cytoplasm.** The complex of chemicals and structures exterior to the cell nuclus (in a eukaryotic cell). This is the space in which chemical reactions occur. Both eukaryotic and prokaryotic cells have cytoplasm. Cytoplasm is a semiliquid material enclosed by the plasma membrane.

♦ **Plasma membrane.** A bilayer of phsopholipids and proteins that surrounds the cytoplasm in a cell, called cell membrane in case of a prokaryotic cell.

♦ **Cell wall.** A structure that contains carbohydrates and surrounds algal, fungal, and most procaryotic and plant cells.

Q5
What are organelles?

Q6
What is an enzyme?

A5

An organelle is a specialized compartment in the cytoplasm of a eukaryotic cell with a specific function.

Examples of Organelles:

Centriole. Supports cell division.

Chloroplast. Supports the process of photosynthesis in green plants.

Cytoskeleton. Supports the cell shape and structure.

Endoplasmic reticulum (ER). Responsible for protein synthesis in the cell and transport of material from the cell.

Golgi body. Involved in the modification and sorting of proteins.

Lysosome. Contains enzymes to digest substances.

A6

A kind of protein molecules that works as a catalyst in a chemical reaction, that is, helps bring about a chemical change without changing itself. Almost all reactions (processes) in a biological cell need enzymes to occur at significant speed.

Q7

To communicate with the external environment the cell cytoplasm moves the material through the plasma membrane. What are some of the mechanisms used for this movement?

Q8

What is active transport?

A7

- ◆ **Active transport**
- ◆ **Diffusion**
- ◆ **Endocytosis**
- ◆ **Facilitated diffusion**
- ◆ **Osmosis**

A8

A process in which a membrane-spanning carrier protein moves specific material from an area of lower concetration to an area of higher concentration, and therefore energy is expended in the process.

An example of active transport is the continued transportation of sodium ions out of the nerve cells to prepare the cell for the upcoming impulse.

Q9

What are diffusion and facilitated diffusion?

Q10

What is endocytosis?

A9

Diffusion is a random movement of molecules from an area of higher concentration to an area of lower concentration as a result of continued collision between molecules.

Facilitated diffusion is the movement of molecules across the membrane facilitated by proteins from high concentration to low concentration.

For example, glucose molecules diffuse slowly across a cell membrance when they diffuse unassisted. However, proteins assist the glucose to diffuse into the cell very quickly (facilitated diffusion).

A10

The process in which cells absob material such as proteins from the external environment by engulfing it with their cell membrane. This process is used to consume polar molecules which cannot be easily moved through the hydrophobic plasma membrane.

Q11

What is osmosis?

Q12

What is energy?

A11

Movement of water molecules through a membrance from an area of high concentration to an area of low concentration.

As an example, osmosis enables the plant roots to suck up water form the soil.

A12

Energy is the ability to perform work. For example, the gasoline in your car provides it the energy that enables the car to do work, that is, to move from one point to another. Cells in living organisms need energy to maintain their structure and functionality.

Q13

What is an endergonic reaction?

Q14

What is an exergonic reaction?

A13

Any chemical reaction in which energy is consumed to carry on the reaction. The products have more energy than the reactants.

A14

Any chemical reaction in which energy is released. The products have less energy than the reactants.

Q15

Chemical reactions in living cells are catalyzed by what?

Q16

What is metabolic pathway?

A15

Enzymes

There are thousands of different kinds of enzymes in a cell catalyzing thousands of different chemical reactions.

For xample, the enzyme that breaks down hydrogen peroxide to water and hydrogen is called *catalase*. Another enzyme called lactases breaks down lactose to glucose and galactose.

A16

A sequence of chemical reactions occurring in a cell. In each pathway, **a** principle substance is modified by the chemical reactions on the pathway. These reactions are catalyzed (accelerated) by enzymes. There are two kinds of pathways:

Catabolism. A type of pathways that break down large complex molecules into smaller units and often release energy. For example large mulecules such as nucleic acids, polysaccharides, and proteins are broken down into nucleotides, monosaccharides, and amino acids.

Anabolism. A type of pathways that build large molecules from smaller units. These pathways usually consume energy. Anabolism helps building tissues and organs.

Q17

A molecule that provides energy for metabolic pathways (or metabolism) is called what?

Q18

What is ATP?

A17

Adenosine triphosphate (ATP)

A18

ATP is a molecule in living cells that provides most of the energy used in metabolism. It's a multifuntional nucleotide that is mostly known as the molecular currency for intercellular energy transfer. Energy released in the catabolic pahways is stored in ATP molecules and is consumed in anabolic pathways.

Q19

How is ATP produced in the cell?

Q20

What is proton gradient in the ATP generation process?

A19

ATP is produced from adenosine diphosphate (ADP) and phosphate ions by using cofactors. In these reactions, a cofactor, which is a non-protein chemical compound tightly bound to an enzyme, acts as a catalyst. These cofactors are also called coenzymes. Here are some examples:

♦ **Flavin adenine dinucleotide (FAD)**

♦ **Nicotinamide adenine dinucleotide (NAD)**

♦ **Nicotinamide adenine dinucleotide phosphate (NADP)**

A20

A buildup of positive charge due to H+ ions, that is, a higher concentration of protons outside the inner membrane of a mitochondrion than inside the membrane.

A mitochondrion is an organelle in the cytoplasm of a cell and is enclosed by double membrane. It functions in energy production.

Q21
What is chemiosmosis?

Q22
What is cellular respiration?

A21

Refers to the use of proton gradient across a membrane to generate cellular energy in form of ATP.

A22

The process of converting chemical energy into cellular energy in form of ATP. There are two types of cellular respiration:

Aerobic. Uses oxygen.

Anaerobic. Does not use oxygen. Fermentation is an example of anaerobic respiration.

Here is an axample of aerobic respiration:

$C_6H_{12}O_6$ + + 38 ADP + 6 O_2 + 38 P

---> 6CO_2 + 38 ATP +6 H_2O + Heat

Q23

What are the main steps in the process of cellular respiration?

Q24

What is photosynthesis?

A23

1. **Glycosis.** Glucose mlecules are broken down to form molecules of pyruvic acid.

2. **Krebs cycle.** Pyruvic acid moleculs are used to form high energy compunds such as nicotinamide adenine dinucleotide with added hydrogen (NADH).

3. **Electron transportation.** Electrons are transported in the chemical reactions. Energy released by the electrons pushes the protons through the membrane to build proton gradient.

4. **Chemiosmosis.** The proton gradient is used for ATP synthesis.

A24

A biochemical process in which the light energy is converted to chemical energy, which is used in synthesizing carbohydrates. Plants use this process by using Sun as the light energy source and carbon dioxide and water as raw materials (reactants) for the process. The end product includes glucose and oxygen.

Q25

What is the chemical equation for photosynthesis?

Q26

In the leaves of plants, photosynthesis occurs in which organelle?

A25

$6CO_2 + 12H_2O +$ Photons (Light energy) $\rightarrow C_6H_{12}O_6 + 6O_2 + 6H_2O$

This equation is sometimes also written in its simplified form:

$6CO_2 + 6H_2O +$ Photons (Light energy) $\rightarrow C_6H_{12}O_6 + 6O_2$

A26

Chloroplast

Q27

What is a photosystem?

Q28

What is glycolysis?

A27

A photosystem is a group of pigment molecules that captures sunlight in the chloroplast. A pigment is the material that absorbs light of certain wavelengths and reflect light of other wavelengths. Therefore it appears colored when sunlight falls on it. Chlorophyll, an example of pigments, is a part of photosystem.

A28

A multistep process in which a glucose molecule is broken down into two molecules of pyruvic acid. It's the first state of cellular aspiration in all organisms, and occurs in the cytoplasm.

Q29

What is Kerb's cycle?

Q30

This is an organelle in eukaryotic cells that receives, modifies, stores, and distributes chemical products of the cell:

A29

The metabolic cycle fueled by acetyl coenzyme that is formed after the process of glycolysis during cellular respiration. The chemical reactions in this cycle complete the breakdown of glucose molecules to cabon dioxide. This cycle supplies most of the NADH molecules to carry energy to the electron transport chain. The Kerb's cycle is also called citric acid cycle.

A30

Golgi apparatus

Q1

What is cell cycle?

Q2

What are the two main phases of a cell cycle?

A1

An orderly series of events that occur in a eukaryotic cell to divide and reproduce. The cycle for each daughter cell extends from the time when the parent cell divides to generate two daughter cells to the time those two daughter cells divide again.

A2

♦ **Interphase**

♦ **Mitotic phase**

Q3

What is interphase?

Q4

What is mitotic phase?

A3

The phase of cell cycle during which the cell performs its normal functions (work) in the organism for which it was created. During this phase the following happens:

♦ The cell approximately doubles everything in its cytoplasm.

♦ The cell increases the supply of proteins.

♦ The cell increases the number of many of its organelles such as mitochondria and ribosomes.

♦ The cell grows in size.

This stage spends typically 90 % of the total cell cycle time.

A4

The phase of the cell cycle during which the cell divides itself into two daughter cells. It includes the following two overlapping processes:

Mitosis. The nucleus and its content are divided and evenly distributed to two daughter nuclei.

Cytokinesis. The cytoplasm is divided. This process usually begins before mitosis ends.

Q5

What is a chromosome?

Q6

What is chromatin?

A5

A thread like linear strand that consists of a DNA molecule and the associated protein that serves to package and manage the DNA. In other words, a chromosome is a physically organized (or packaged) unit of DNA in a cell.

A6

The substance distributed in the nucleus of a cell that condenses into chromosomes during cell division. It consists of DNA and protein.

Q7

What are nucleosomes?

Q8

What are histones?

A7

Repeating subunits of chromatin

A8

Small protein molecules found in chromatin in association with DNA. They play role in packaging DNA in the eukaryotic chromosome.

Q9

What are the different phases of the mitosis process?

Q10

What is prophase in mitosis?

A9

- ◆ **Prophase**
- ◆ **Metaphase**
- ◆ **Anaphase**
- ◆ **Telophase**

A10

The first stage of mitosis in which the nuclear envelop breaks down and chromatin condenses to form chromosomes

Q11

What is metaphase?

Q12

What is anaphase?

A11

The second stage of mitosis during which the duplicated chromosomes of the cell are lined up along the center of the cell called the equatorial plate or metaphase plate

A12

The third stage of mitosis that begins with the daughter chromatids separating from each other and ends with a complete set of sibling chromosomes arriving at each of the two poles of the cell

Cell Cycle

Q13

What is telophase?

Q14

Which cells in human body do not divide?

A13

The fourth and the last stage of mitosis during which the daughter nuclei form at the two poles of the dividing cell

A14

Red blood cells

Q15

What is the main function of red blood cells?

Q16

What is meiosis?

A15

Deliver oxygen from the lungs or gills to the tissues

A16

The process in which a diploid cell of a sexually reproducing organism divides into four haploid daughter cells, also called gametes. These cells (gametes) are produced in the sex organs of parents.

Q17

What is a diploid cell?

Q18

What is a gamete?

A17

A cell that contains two sets (homologous pairs) of chromosomes, one set inherited from each parent. For example, in a human cell, 46 chromosomes are organized into 23 pairs.

A18

A sex cell: a haploid egg (produced by a female) or sperm (produced by a male). A gamete, say sperm, from one parent fuses with a gamete, say egg, from the other parent during fertilization.

Q19

What are two main steps in meiosis?

Q20

Red blood cells contain how many chromosomes?

A19

1. A cell duplicates itself. For example, a human cell with 46 chromosomes divides itself into two daughter cells each with 46 chromosomes.

2. Each daughter cell divides itself into two cells each with 23 chromosomes.

A20

0

Q1
What is Mendelian genetics?

Q2
Did Mendel know of DNA molecules?

A1

The theory of inheritance of traits (characteristics) based on the work performed by Gregor Mendel during 1860s and 1870s. It's also called classical genetics to distinguish it from what is called molecular genetics.

A2

No

Q3

Mendel developed a method to predict what?

Q4

What are inheritance patterns?

A3

The outcome of inheritance patterns

A4

Patterns observed in the transmission of traits (or characteristics) from one generation to the next

Q5

Mendel performed experiments on which organisms?

Q6

Mendel fertilized seeds from a line with round peas with pollen from the line of wrinkled peas. What was the offspring?

A5

Pea plants

A6

Round peas

Q7

Mendel fertilized round peas with wrinkled peas and the offspring was round peas. Then he self-fertilized these round peas, that is, he put pollen of a round pea plant to the egg of the same plant. What was the offspring?

Q8

How can you explain the results of Mendel's experiments?

A7

Three round peas to one wrinkled pea

A8

When round peas mixed with the wrinkled peas, each combination had a gene from the wrinkled pea and a gene from the round pea. The round gene was the dominant gene. So, all peas looked round.

During self-fertilization, each interacting pea had one wrinkled gene and one round gene. So the output had one pea that had both wrinkled genes for three peas that had at least one round gene. Because the round gene is the dominant gene and hides the wrinkled gene, that resulted in one wrinkled pea for three round peas.

Q9

Sate some of Mendel's laws of genetics.

Q10

What is an allele?

A9

Law of dominance. When there are two different alleles for a trait of an organism, the dominant allele hides the recessive allele.

Law of separation. During the formation of gamete from diploid (as in meiosis), the members of a pair of alleles for a particular trait separate from each other.

Law of independent assortment. The separation of members of a gene pair from each other during meiosis happens independent of the members of other gene pairs.

A10

A member of a pair or a series of genes that occupy a specific position on a specific chromosome. Some alleles are dominant, whereas others are recessive.

Q11

What is genetic cross?

Q12

What is a Punnett square?

A11

A process to predict the traits of the offspring

A12

A diagram used by biologists to determine the probability of the traits of an offspring given the genotype (genetic constitution) of the parents

Q13

Consider a male and a female both with genotypes Bb, where B is a dominant allele and b is a recessive allele. Draw the Punnett square for the possible genotypes of their offspring.

Q14

How many sex chromosomes there are in a human cell?

A13

Male alleles (B and b) combining with female alleles (B and b) to form different genotypes for the offspring:

	Male B	Male b
Female B	BB	bB
Female b	Bb	bb

Probability for BB: 25%

Probability for bb: 25%

Probability for Bb or bB: 50%

A14

Out of 23 pairs of chromosomes in a human cell, one pair is the sex chromosomes pair.

Q15

The sex chromosomes in human cells determine what?

Q16

What is a genotype?

A15

The sex (gender) of humans

A16

The gene composition of a living organism, that is, the specific allelic makeup.

Q17

What is a phenotype?

Q18

What are the types of sex chromosomes?

A17

The physical appearance of an organism that results from the interaction between its genetic makeup and the environment:

genotype + environment --> phenotype

A18

There are two types of sex chromosomes: X and Y.

Q19

What are the sex chromosomes pattern of males and females?

Q20

Assume that the gene that determines the hair color lives on the X chromosome. Use the Punnett square to determine the hair color of sons and daughter of a female with white hair and a male with black hair. Further assume that the black hair gene is dominant over the white hair gene.

A19

Female has both chromosomes in the pair as X: XX

Male has one X chromosome and one Y; therefore the pattern is: XY

A20

Male alleles (X and Y) combining with female alleles (X and X) to form different genotypes for the offspring:

	Male X_{black}	**Male Y**
Female X_{white}	$X_{black} X_{white}$ (daughter)	YX_{white} (son)
Female X_{white}	$X_{black} X_{white}$ (daughter)	$Y X_{white}$ (son)

Hair color of sons will be white.

Hair color of daughters will be black because black hair gene is dominant over white hair gene.

Q1

What is genetics?

Q2

What is a gene expression?

A1

The science of heredity and hereditary variations. In other words, it's a study of genes and how they are expressed and inherited.

A2

Gene expression is the process in which an RNA and then a protein is synthesized according to the information coded in the gene.

Q3

Molecular genetics deals with the activity of which molecule?

Q4

What is DNA?

A3

DNA

A4

Deoxyribonucleic acid; the genetic material that organisms inherit from their parents and that contains the instructions to make protein.

Q5

What is the structure of a DNA molecule?

Q6

What in a DNA molecule determines the genetic information?

A5

Double helix. The two nucleotide chains twist around each other to make this double helix shape.

A6

The sequence of nucleotides in the nucleotide chains

Q7

What is the central dogma of molecular biology?

Q8

What is the genome of an organism?

A7

The central dogma of molecular biology describes how the genetic information is used in synthesizing proteins:

1. **DNA Replication.** DNA is copied to DNA for inheritance.

2. **Transcription.** The information contained in a part of DNA is transferred to newly created messenger RNA (mRNA).

3. **Translation.** The mRNA moves into ribosome where it is translated to synthesize protein.

A8

The entire sum of DNA in the cell including genes and non-coding sequences.

Q9

DNA and RNA are two types of nucleic acids.

What is a nucleic acid?

Q10

What is a nucleotide?

A9

A nucleic acid is a polymer composed of monomers called nucleotides.

A10

A nucleotide is a compound molecule with the following structure:

♦ A five-carbon sugar at the center: deoxyribose in DNA and ribose in RNA

♦ Negatively charged phosphate group (PO_4^-) attached to the sugar

♦ A base containing nitrogen called nitrogenous base attached to the sugar

Each DNA molecule is composed of two strands, and each strand is a polynucleotide composed of four different kinds of nucleotides.

Q11

What are the four different kinds of nucleotides in DNA?

Q12

What keeps the two strands of a DNA molecule together?

A11

Each nucleotide in DNA has the same sugar and phosphate group. So, the four kinds of nucleotides are distinguished by the four different nitrogenous bases. Each nucleotide contains one of the following four nitrogenous bases:

- ♦ **Adenine (A).** A double ring structure of carbon, hydrogen, and nitrogen atoms.

- ♦ **Guanine (G).** A double ring structure of carbon, hydrogen, and nitrogen atoms.

- ♦ **Cytosine (C).** A single ring structure of carbon, hydrogen, nitrogen, and oxygen atoms.

- ♦ **Thymine (T).** A single ring structure of carbon, hydrogen, nitrogen, and oxygen atoms.

A12

The hydrogen bonds (H-bonds) between the nitrogenous bases of the two strands. Only two base pairings are possible:

- ♦ A-T

- ♦ C-G

Q13
Where in the body is DNA made?

Q14
Where does DNA reside in a prokaryotic cell?

A13

DNA is made in cells.

A14

Floating in the cytoplasm

Q15

What is a plasmid?

Q16

What are vectors?

A15

A ring shaped DNA (separate from the chromosomal DNA) found in prokaryotes and yeasts

A16

The plasmids used in genetic engineering to transfer genes from one cell to another

Q17

What are R plasmids?

Q18

What is biotechnology?

A17

Plasmids that contain antibiotic resistance genes

A18

According to the United Nations Convention on Biological Diversity:
"Biotechnology means any technological application that uses biological systems, living organisms, or derivatives thereof, to make or modify products or processes for specific use."
Biotechnology can also be defined as a study and manipulation of living organisms or their components such as organs, tissues, cells, and molecules, for different purposes such as performing research and developing products and applications.

Q19

Researchers in biotechnology use knowledge, tools, techniques, and processes from which fields?

Q20

List some techniques and practices of biotechnology.

A19

- **Biology**
- **Computer science**
- **Chemistry**
- **Physics**

A20

- **Cloning**
- **Fermentation**
- **Recombinant DNA**

Q21

What is cloning?

Q22

What is fermentation?

A21

The technique (or process) to generate a genetically identical copy of an organism, cell, or a DNA fragment (molecule).
For example, human cloning refers to creating a new human who will be a genetically identical copy of a living (or a dead) human.
Dolly the sheep (07/05/1996 – 02/14/2003) was the first mammal that was cloned from an adult cell at Roslin Institute in Scotland. Her birth was announced on February 1997.

A22

A technique or process to create energy in a cell by converting sugar into lactic acid or ethanol. Sugars are the common substrate (initial substance or reactants) of fermentation.
A common example:
Yeast is used as a catalyst to carry out fermentation in order to produce ethanol in alcoholic drinks such as beer or wine.

Q23

What is recombinant DNA technique?

Q24

What is human insulin?

A23

A technique for splitting and recombining the DNA molecules. A recombinant DNA molecule is a DNA molecule that contains DNA from at least two different sources.
This ability to cut and paste DNA enables biotechnology companies to develop a wide variety of products and applications. Human insulin is an example.

A24

A protein (polypeptide hormone) that regulates the metabolism by faciliatating uptake of sugar from blood into cells. Genetically engineered insulins are used for the treatment of diabetes.

Q25

What is diabetes?

Q26

What are antibiotics?

A25

A disorder that affects the uptake of sugar from blood into cells resulting in high sugar in the blood

A26

Chemical substances produced by a few species of bacteria or fungi that kills some other microrganisms. The first antibiotic product developed as medicine was penicillin to treat infections.

Q27

What is the basis of DNA fingerprinting?

Q28

List some categories of products that most of the biotechnology companies develop?

A27

The genomic loci and the length of certain types of small repetitive sequences vary significantly from human to human.

A28

♦ **Agricultural**

♦ **Industrial**

♦ **Pharmaceutical**

♦ **Research or production tools**

Q29

What is a virus?

Q30

How are viruses used in biotechnology?

A29

A nano (or sub-microscopic) particle that has the capability of infecting cells of living organisms. A virus contains genetic material protected in a protein coat. Viruses usually range from 20 to 300 nm in size. Unlike cells, viruses cannot multiply on their own, and can only replicate themselves by infecting a host cell.

A30

♦ **Because viruses infect the living organisms, they are often the target of the biotechnology solutions such as products and therapies.**

♦ **Virus particles are used in biotechnology research to carry DNA from one cell to another.**

Q31

What is gene therapy?

Q32

What is genetic engineering?

A31

The treatment of a disease by providing the patient with a new gene, for example, replacing dysfunctional genes with functional genes. It's also called human gene therapy.

A32

The scientific way of altering genes or genetic materials for the purpose of either creating new desirable traits in organisms or to eliminate the undesirable traits

Q33

What is the general process of genetic engineering?

Q34

What is gene regulation?

A33

1. Identify the molecules that need to be produced through genetic engineering.

2. Isolate the DNA (gene) that contains the instructions to produce the identified molecules.

3. Manipulate the DNA instructions to produce the desirable product. This can be accomplished by one of the following two ways:

   o Change the DNA instructions inside the cells of the target organism.

   o Introduce the new instructions into the cells of another organism that can produce the desired molecules more efficiently.

4. Manufacture the molecules (product).

5. Test the molecules for their new traits.

A34

The process of turning specific genes on and off within a living organism

Q35

What is mutation?

Q36

This is the process used to amplify the production of copies of a DNA:

A35

A change in the nucleotide sequence of DNA; this is the ultimate source of genetic diversity.

A36

PCR; Polymerase Chain Reaction

Q1

What is evolution of life?

Q2

Who is the author of Origin of Species?

A1

The process through which a population changes one or more characteristics (traits) over time. This comes through many changes including genetic changes in a population or species that occur over generations, heritable changes that produce diversity of life, and all other changes that transform life.

A2

Charles Darwin

Q3

According to Darwin, what is the mechanism for evolution of life?

Q4

What are the two essential points of the theory of natural selection?

A3

Natural selection

A4

♦ **Genetic changes within a population occur in a random fashion, and lead to gradual development of a population or a species over generations, for example, giving rise to new traits.**

♦ **In the given environment of a population, the fittest survive and spread their traits through the population.**

Q5

True or false: Evolution occurs in individuals.

Q6

What is a population?

A5

False. Evolution occurs in populations; populations are the units of evolution. In other words, a population is the smallest biological entity (unit) that can evolve.

A6

Group of individuals of the same species living in the same place at the same time

Q7

What is a species?

Q8

Variations are the raw material for evolution. What branch of biology explains these variations?

A7

A group of organisms whose members possess similar anatomical characteristics and have the capability to interbreed. A species may have multiple populations.

A8

Molecular biology or molecular genetics

Q9

Natural selection is the mechanism for evolution described in the Origin of Species. What are some other mechanisms for evolutions, the mechanisms that work at gene level?

Q10

What is a gene pool?

A9

◆ **Gene flow**. Refers to the flow of genes from one population to another. For example, a population may gain (or lose) alleles when fertile individuals move into (or out of) the population. Other example is the transfer of gametes between two populations.

◆ **Genetic drift**. A statistical change in the gene pool of a population.

◆ **Mutation**. A change in the nucleotide sequence of DNA; this is the ultimate source of genetic diversity.

A10

A conceptual collection of all the genes of a population at a given time. It consists of all the alleles in the population at a given time, and serves as a reservoir from which the next generation (offspring) draws its genes.

Q11

True or false: All variations in a population are heritable.

Q12

List some evidences for evolution.

A11

False. Not all variations are heritable.

A12

♦ **The fossil records**

♦ **Biogeography.** The geographic distribution of species.

♦ **Comparative anatomy**

♦ **Comparative embryology**

♦ **Molecular biology and biochemistry**

Q13

What are the two major branches of prokaryotic evolution?

Q14

The bacteria that use photosynthesis to produce carbohydrates are called:

A13

- **Archaea**
- **Bacteria**

A14

Cyanobacteria

Evolution of Life

Q15

Living things first came from the ocean to occupy the land during which era?

Q16

During which era reptiles such as dinosaurs evolved and became the predominant organisms on Earth?

A15

Paleozoic era

A16

Mesozoic era

Q17

At the end of which era, the great extinction took place in which dinosaurs disappeared?

Q18

Which era is called the age of mammals?

A17
Mesozoic era

A18
Cenozoic era

Q19

What is the name of the group to which the first hominids (human like creatures) belonged?

Q20

We, the today's human, are variations of which group?

A19

Australopithecus

A20

Homo sapiens; means intelligent human.

Q21

Homo sapiens are believed to be evolved from:

Q22

Homo erectus replaced:

A21

Homo erectus

A22

Homo habilis

Q23

What was the predominant species during the stone age?

Q24

The first hominids that were able to walk upright are called what?

A23

Homo habilis

A24

Australopithecus

Q25

Which group of hominids is considered to be the first human?

Q26

The concept of language first existed in which of the hominid group?

A25

Homo habilis

A26

Homo erectus

Q27

Biologists classify the diversity of life into three main groups called domains. Two of these domains are bacteria and archaea. What is the third domain called?

Q28

The domain eukarya is classified into at least four smaller categories called kingdoms. Two of these kingdoms are fungi and plants. What are the other two kingdoms?

A27

Eukarya

A28

Animalia and protists

Q29

The branch of biology that deals with identifying, naming, and classifying species of organisms is called:

Q30

What accounts for the diversity of life on Earth?

A29

Taxonomy

A30

Evolution

Q31

What is the relationship between types of cells, domains of life, and kingdoms of life?

Q32

The wind carrying seeds to a new (uninhabited) site is an example of:

A31

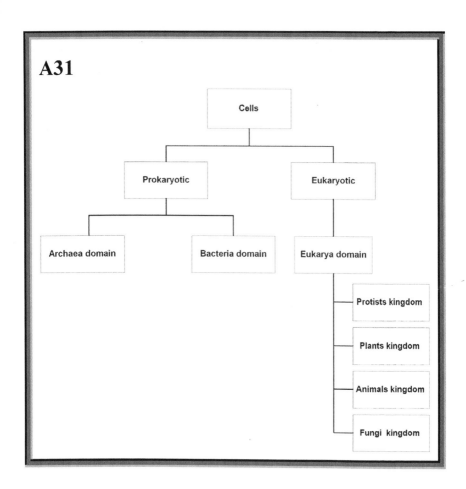

A32

Gene flow

Q1

True or false: Most prokaryotes are unicellular.

Q2

The two domains of prokaryotic evolution are archaea and:

A1

True

A2

Bacteria

Q3

Spherical species of bacteria is called:

Q4

The rod-shaped species of bacteria is called:

A3

Cocci (singular; coccus)

A4

Bacilli (singular; bacillus)

Q5

The spiral shaped bacteria are called:

Q6

Most bacteria acquire their food from organic matter. For that reason, they are called:

A5

Spirochetes (rigid) and spirilla (flexible).

Singular for spirilla is spirillum.

A6

Heterotrophic

Q7

A large number of bacteria feed on dead or decaying organic matter. For that reason, they are called:

Q8

Some bacteria synthesize their own food. For that reason, they are called:

A7

Saprobic

A8

Autotrophic

Q9

These autotrophs harness light energy to synthesize organic compounds from CO_2:

Q10

These autotrophs use chemical reactions as a source of energy:

A9

Photosynthetic or photoautotrophs

A10

Chemosynthetic

Q11

Bacteria living at very cold temperatures are called:

Q12

Bacteria living at the human body temperatures are called:

A11

Psychrophilic

A12

Mesophilic

Q13

Bacteria living at very high temperatures are called:

Q14

Bacteria that require molecular oxygen (O_2) for their metabolism are called:

A13

Thermophilic

A14

Aerobic

Q15

Bacteria that don't require molecular oxygen (O₂) are called:

Q16

These bacteria can live with or without oxygen:

A15

Anaerobic

A16

Facultative. They make energy through aerobic respiration if oxygen is available, and switch to fermentation if oxygen is not available.

Q17

True or false: Many bacteria play a very useful role beneficial to the environment.

Q18

True or false: In the food industry, some bacteria are used to prepare products such as cheese, dairy products, and pickles.

A17

True

A18

True

Q19

Some bacteria cause diseases, and they are called:

Q20

Poisonous proteins secreted by bacterial cells are called:

A19

Pathogens

A20

Exotoxins

Q21

The poisonous chemical components of the cell walls of some pathogen bacteria are called:

Q22

About 2 billion years ago, THESE bacteria enriched the earth's environment with the oxygen they produced:

A21

Endotoxins

A22

Cyanobacteria. These bacteria are photosynthetic

Q23

This is a microscopic particle capable of infecting cells of living organisms by inserting its genetic material into the cells:

Q24

Viruses that attack bacteria are called:

A23

Virus

A24

Bacteriophages

Q25

True or false: A Virus is made of cells.

Q26

True or false: A virus can make more viruses only by infecting a cell with its genes and then using the molecular machinery of the cell to replicate itself.

A25

False. A virus does not have cells. Yes, all living organisms are made of cells, but viruses are not considered living organisms. They are life-like though, but most of them only have genes and cannot reproduce on their own. Therefore, in most cases, a virus is nothing more than genes in a box.

A26

True

Q27

The first eukaryotes evolved from the prokaryotes were:

Q28

True or false: Most of the protists are unicellular.

A27

Protists

A28

True

Q29

The eukaryotes that do not belong to any of the three kingdoms, plants, animals, and fungi, belong to the kingdom called:

Q30

The protists that live primarily by ingesting food are called:

A29

Protists

A30

Protozoans

Q31

Most species of *THESE protozoans* move and feed by means of temporary extension of their cell, called pseudopodia:

Q32

These protists resemble fungi:

A31

Amoebas

A32

Slime molds

Q33

Photosynthetic protists are called:

Q34

True or false: Prokaryotes contain genetic (hereditary) material in the nucleus of the cell.

A33

Algae (singular; alga)

A34

False

Q1

What are fungi?

Q2

What is a fungus?

A1

A kingdom of eukaryotes

A2

A eukaryote that digests its food externally by absorbing small nutrient molecules from the surrounding medium

Q3

The structure of fungi is adaptive to externally absorb nutrients, and this structure is called:

Q4

Fungi reproduce by releasing:

A3

Hypha (plural; hyphae)

A4

Spores. These spores are produced either sexually or asexually.

Q5

True or false: Like bacteria and protists, fungi are the major decomposers of organic matter on earth.

Q6

True or false: Some fungi are parasitic and cause plant diseases.

A5

True

A6

True

Q7

The first antibiotic, penicillin, was made from the common mold called:

Q8

The Eumycota division of fungi is subdivided into five classes. Name those classes:

A7

Penicillium

A8

- ◆ Ascomycetes
- ◆ Basidiomycetes
- ◆ Deuteromycetes
- ◆ Oomycetes
- ◆ Zygomycetes

Q9

Yeasts are unicellular and belong to which class of fungi?

Q10

Common mushrooms belong to which class of fungi?

A9

Ascomycetes

A10

Basidiomycetes

Q11

This class of fungi only reproduces through an asexual process:

Q12

Water molds belong to which class of fungi?

A11

Deuteromycetes

A12

Oomycetes

Q13

Bread mold belongs to which class of fungi?

Q14

A mutualistic association between a fungus and a cyanobacterium or between a fungus and an alga is called:

A13

Zygomycetes

A14

Lichen

Q15

What is an alga?

Q16

True or false: Many fungi are parasites on animals (including human), plants, and other fungi.

A15

A variety of protists

A16

True

Q17

True or false: Fungi cause many agricultural and human diseases.

Q18

A living association between two organisms is called:

A17

True

A18

Mutualism

Q1
What is a plant?

Q2
Plants are considered to be originated from their protists ancestors called:

A1

A plant is a multicellular eukaryote that creates organic molecules through photosynthesis. In other words, plants produce their own food through photosynthesis.

A2

Green algae or charophyceans

Q3

These plants contain two types of specialized tissues called phloem and xylem:

Q4

The plant tissue that gets water and nutrients from the roots to the rest of the plant is called:

A3

Vascular plants

A4

Xylem

Q5

The tissue that transports sugar and other nutrients from the leaves to other parts of the plant is called:

Q6

True or false: Ferns are seedless vascular plants.

A5

Phloem

A6

True

Q7

A nonvascular plant, that is, a plant that lacks xylem and phloem is called:

Q8

True or false: Mosses are vascular plants.

A7

Bryophyte

A8

False. Mosses are bryophytes

Q9

The vascular plants with unprotected (naked) seeds are called:

Q10

The vascular plants with protected (contained) seeds are called:

A9

Gymnosperms

A10

Angiosperms

Q11

The cone-bearing gymnosperms are called:

Q12

The two most distinguishing features of angiosperms are:

A11

Conifers

A12

Flower and fruit

Q13

The flower of an angiosperm consists of a ring of modified leaves that encloses and protects the flower bud before it opens. These modified leaves are called:

Q14

True or false: A pine tree is a conifer that produces male and female cones on the same tree.

A13

Sepals

A14

True

Q15

This nutrient-rich mass is formed during fertilization and provides nourishment to the developing embryo in the seed:

Q16

This plant structure anchors the plant to the ground, and absorbs and transports minerals and water:

A15

Endosperm

A16

Root

Q17

This part of a plant's shoot system supports the leaves and reproductive structures:

Q18

The principal organs to support photosynthesis in vascular plants are called:

A17
Stem

A18
Leaves

Q19

True or false: The development and growth of many plants are regulated by the plant hormones.

Q20

The interaction of the plant hormones with the stimuli from the environment can trigger a bending or turning of the plant. This response from the plant is called:

A19

True

A20

Tropism

Q21

The bending of the plant toward the light is called:

Q22

The turning of a plant away from or toward the earth is called:

A21

Phototropism

A22

Geotropism

Q23

True or false: The flow of water from the roots to the leaves depends on adhesion, air pressure, cohesion, and humidity.

Q24

The wood of the stem is developed from which tissue?

A23

True

A24

Xylem

Q25

Which plant tissue combined with the tough tissue called cork becomes the bark of the stem?

Q26

This is the plant hormone that inhibits the growth of developing leaves:

A25

Phloem

A26

Abscisic acid

Q1

What is an animal?

Q2

What is a heterotrophic organism?

A1

A eukaryotic, multicellular, heterotrophic organism that takes food by ingestion, that is, take in food and digest it into smaller pieces

A2

An organism that cannot make its own food (organic molecules), and must obtain it by consuming organic products or by consuming other organisms.

Q3

What are the two major groups of animals?

Q4

Animals with backbones are called:

A3

- ◆ **Vertebrate**
- ◆ **Invertebrate**

A4

Vertebrates

Q5

Animals that have no backbones are called:

Q6

What is phylum?

A5

Invertebrates

A6

The taxonomic category below kingdom but above class

(plural, *phyla*)

Q7

List some of the phyla among invertebrates.

Q8

Sponges belong to the phylum called:

A7

- Annelida
- Arthropoda
- Chordata
- Cnidaria
- Echinodermata
- Mollusca
- Nematoda
- Porifera

A8

Porifera

Q9

Earthworms and leeches belong to the phylum called:

Q10

Insects, lobsters, and spiders belong to the phylum called:

A9
Annelida

A10
Arthropoda

Q11

The largest phylum in the animal kingdom is:

Q12

All members of this phylum has a bilateral symmetry and include both vertebrates and invertebrates:

A11

Arthropoda

A12

Chordata

Q13

Jelly fish, sea anemones, and sea corals belong to the phylum:

Q14

Sea urchin and starfish belong to the phylum:

A13

Cnidaria

A14

Echinodermata

Q15

Snails, octopus, and a variety of organisms called seafood belong to the phylum:

Q16

Roundworms make up the phylum:

A15

Mollusca

A16

Nematoda

Q17

All echinoderms contain an internal support system, a hard skeleton, called:

Q18

List some classes of vertebrates.

A17

Endoskeleton

A18

- ◆ Amphibia
- ◆ Aves
- ◆ Fishes
- ◆ Mammalia
- ◆ Reptiles

Q19

This class was the first vertebrates to colonize land:

Q20

Frogs and toads belong to the class of vertebrates called:

A19

Amphibia

A20

Amphibia

Q21

Birds that belong to the class of vertebrates are called:

Q22

Birds can maintain a constant body temperature, that is, the birds are:

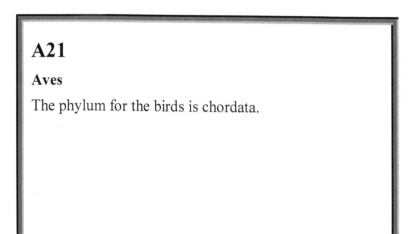

A21

Aves

The phylum for the birds is chordata.

A22

Homeothermic

Q23

The birds use their own metabolic heat to maintain a constant body temperature. That means the birds are:

Q24

This class consists of jawless vertebrates, which are aquatic:

A23

Endotherms

A24

Fish

Q25

Mammals belong to the class of vertebrates called:

Q26

Animals that produce milk with mammary gland belong to the class of vertebrates called:

A25
Mammalia

A26
Mammalia

Q27

Mammals that lay eggs are called:

Q28

Alligators, crocodiles, lizards, snakes, and turtles belong to the class of vertebrates called:

A27

Monotremes

A28

Reptiles

Q29

Fishes exchange gas with the environment by using:

Q30

What is nutrition?

A29

Gills

A30

The process used by living organisms to take in and utilize food material, that is, to obtain the food from the environment and transport it to the cells.

Q31

What is a nutrient?

Q32

List some nutrients used by animals.

A31

Any substance that can be metabolized by a living organism to produce energy

A32

♦ **Carbohydrates**

♦ **Lipids**

♦ **Minerals**

♦ **Nucleic acids**

♦ **Proteins**

Q33

What is digestion?

Q34

What are vitamins?

A33

The process of mechanical and chemical breakdown of food into molecules suitable for the body to absorb

A34

Organic nutrients required by living organisms in small quantities

Q35

What are some of the minerals required by animals?

Q36

The most commonly used carbohydrate as a source of energy is:

A35

♦ **Magnesium**

♦ **Phosphorus**

♦ **Potassium**

♦ **Sulfur**

♦ **Zinc**

A36

Glucose

Q37

From where do the animals get nucleic acids?

Q38

The spontaneous movement of particles (such as molecules) from a region of higher concentration to a region of lower concentration:

A37

Plants and animal tissues

A38

Diffusion

Q39

The tendency of the body of a living organism to maintain constant conditions (steady state) in its internal environment even when the external environment changes:

Q40

The fundamental functional and structural unit of a nervous system is called:

A39

Homeostasis

A40

Neuron, the nerve cell

Q41

What is a nerve?

Q42

A system of organs that forms a coordination and communication network in an animal's body:

A41

A cable like bundle of neuron fibers that are tightly wrapped in a connective tissue

A42

Nervous system

Q43

List some organs of human senses.

Q44

This hormone promotes breast development and milk secretion in females:

A43

- ◆ **Ear**
- ◆ **Eye**
- ◆ *Chemoreceptors* **of human tongue (taste)**
- ◆ *Chemoreceptors* **of human nose (smell)**
- ◆ *Pacinian corpuscles* **in the skin (touch and pain)**

A44

Lactogenic hormone

Q1
What is ecology?

Q2
What are some of the levels of living organisms?

A1

The scientific study of interaction of living organisms with their environment

A2

- ◆ Population
- ◆ Community
- ◆ Ecosystem
- ◆ Biosphere

Q3

What is population?

Q4

What is a community?

A3

A group of living organisms of the same species living in the same geographic area

A4

An assemblage of organisms possibly belonging to different species living in the same area and potentially interacting with each other

Q5
What is an ecosystem?

Q6
What is the biosphere?

A5

A biological community along with its environment, that is, all the living organisms in an area along with the environment that they interact with

A6

The portion of earth that is alive, that is, the global ecosystem

Q7

What is population density?

Q8

What is population ecology?

A7

The number of living organisms (individuals) of a species per unit area or per unit volume

A8

The study of how members of a population interact with their environment

Q9

An interaction between two or more species (populations) that live together in a community in direct physical contact is called:

Q10

The symbiotic relationship in which one species called the parasite benefits by harming its host:

A9

Symbiotic relationship

A10

Parasitism

Q11

The symbiotic relationship in which both involved species benefit:

Q12

The symbiotic relationship in which one species participating in the relationship benefits, while the other species neither benefits nor harmed:

A11

Mutualism

A12

Commensalism

Q13

The symbiotic relationship in which two populations accomplish together something that neither of them could accomplish on its own:

Q14

The interaction between two species in which one species, the predator, eats the other, the prey:

A13

Synergism

A14

Predation

Q15

The maximum size of a population that the environment can sustain:

Q16

What is a producer?

A15

Carrying capacity

A16

An organism such as plant, alga, or autotrophic bacterium that makes organic molecules from CO_2, H_2O, and other inorganic raw material. Producers trap (store) energy in an ecosystem.

Q17

What are consumers?

Q18

A sequence of food transfers from producers to consumers is called:

A17

The organisms that feed on the producers

A18

Food chain

Q19

A network of interconnecting food chains is called:

Q20

What is biomass?

A19

Food web

A20

The total mass (or amount) of organic material in an ecosystem

Q21

Major types of ecosystems that cover large geographic regions, such as deserts and tropical forests:

Q22

Passage of energy through the components of ecosystem is called:

A21

Biomes

A22

Energy flow

Q23

True or false: Energy flow and chemical cycling are two major processes that sustain ecosystems.

Q24

What is chemical cycling?

A23

True

A24

The use and reuse of chemical elements such as carbon and nitrogen within an ecosystem